The Red Crossbills of Colorado

Alfred A. Bailey

DENVER MUSEUM OF
NATURE & SCIENCE

Expand your knowledge.
Increase your power.

BCR's Shelf2Life program grew from a strong desire to help libraries share their vital collections with new audiences. BCR is the nation's oldest and most established library network, and their mission is to bring libraries together for greater success by expanding their knowledge, reach and power. By helping libraries digitize and widen access to their collections, BCR's Shelf2Life program helps libraries increase the visibility, use and recognition of their important collections. In addition, the program is designed to help researchers, collectors and all curious readers by putting these editions within easy reach: in print-on-demand or electronic format.

DENVER MUSEUM OF NATURE & SCIENCE

The mission of the Denver Museum of Nature & Science is to inspire curiosity and excite minds of all ages through scientific discovery and the presentation and preservation of the world's unique treasures. The vision of the Museum is to create a community of critical thinkers who understand the lessons of the past and act as responsible stewards of the future. In order to continue to meet this mission and vision the Denver Museum of Nature & Science continually strives to innovate and adapt their collections and learning opportunities. By participating in the Shelf2Life Program, the Museum provides their community access to important knowledge and content in ways they have never been able to in the past.

Loxia curvirostra neogaea Griscom

A medium-sized crossbill with a bill of medium length, but relatively slender, the tip of the upper mandible greatly prolonged beyond the end of the lower; wing 86.5-91; culmen 15.5-17.5; depth of bill 9-10. *Adult males* chiefly dull brick or dark coral red, the feathers of the mantle with distinctly brownish tips; about 10% of males are brighter, distinctly scarlet in tone in fresh winter plumage; worn post-breeding birds are likely to be more scarlet; xanthochroistic males are rare. *Adult females* are dull olive greenish yellow, the feather of the mantle with sooty grayish tips in fresh plumage; dusky spotting on the chest is of rare occurrence and always slight compared with European races; the gray throat is usually untinged with olive and consequently more contrasted with the color of the chest.

Breeds.—Chiefly the Upper Canadian zone of North America, east of the treeless areas. Breeding birds examined from Quebec, Nova Scotia, New Brunswick, Ontario, northern New England, northern New York, Michigan, Wisconsin, and Minnesota. Unquestionably also breeding in the forested portions of Manitoba and Saskatchewan.

Loxia curvirostra pusilla Gloger

A large crossbill, with a long but relatively slender bill, wing of males 91-96.5; culmen 17-19; depth of bill 10.3-11.5; conspicuously larger than *neogaea*, and both sexes much darker in general coloration, but adult males chiefly brick red in fresh plumage, as in *neogaea;* in size quite indistinguishable from many individuals of two western races, both sexes readily recognizable by their dark, dull coloration.

Breeds.—so far as known in Newfoundland only.

Loxia curvirostra minor Brehm

The smallest of New World crossbills with a very short and stumpy bill; wing (♂), 81-88.5; culmen, 13.5-15; depth of bill, 8.0-8.8; coloration precisely as in *neogaea*, fully adult males dull brick red in fresh plumage; in large series from any one locality, notable for the high percentage of xanthochroistic males and the scarcity of fully developed red ones.

Breeds.—The humid coastal strip of the northwestern Pacific coast district from southern Alaska south along the coast of British Columbia, including the Queen Charlotte and Vancouver Islands, to the coastal ranges of Washington and northwestern Oregon. So far as I know the normal northern limit is Cook Inlet (Osgood).

Loxia curvirostra bendirei Ridgway

Exceedingly close to *neogaea*, some individuals quite indistinguishable; both sexes averaging a little larger, longer winged and with a longer bill; the great majority of adult males bright scarlet, rather than dull brick red; both sexes often with a darker and sootier gray, less brownish mantle in fresh plumage; adult females not otherwise separable in color.

Breeds.—More or less regularly in south-central British Columbia, eastern Washington, eastern Oregon, Idaho, western half of Montana and western Wyoming; north (irregularly?) to northern British Columbia (Telegraph Creek) and wandering casually to Vancouver Island (Parkville, May, 1910) and Alaska, Admiralty Island, May 3, 1907.

Loxia curvirostra benti Griscom

A relatively large crossbill, with a long and relatively slender bill; wing (♂) 93.0-98; culmen 17.0-19.0; depth of bill at base 10.0-10.5. Coloration of adult male chiefly in fresh fall and winter plumage strikingly rosy red, with paler and whiter belly, less brownish gray; in worn breeding plumage always bright scarlet. Adult female in fresh fall and winter plumage a lighter and brighter yellowish below, the throat whiter and less flecked with gray, more sharply contrasted with the yellowish breast; belly whiter and grayer, less brownish gray; worn breeding birds often inseparable from *bendirei* in coloration.

Breeds.—Pine hills of southeastern Montana, eastern Wyoming (Weston and Crook counties), western North and South Dakota, and the Rocky Mountain region of Colorado.

Loxia curvirostra grinnelli Griscom

A large crossbill, the adult male scarlet in general coloration throughout the year; wing length and exposed culmen exactly as in *benti;* differing from *benti* in having a much deeper bill, 10.3-11.5; versus 10.0-10.5, and never having the pronounced rosy and paler coloration of that race; easily separable from *bendirei* in much larger size and deeper bill. Also readily separable from *stricklandi*, a still larger bird, with a still deeper bill, and darker coloration, in adult males blood red rather than scarlet.

Breeds.—Mountains of California and northern Arizona.

Loxia curvirostra stricklandi Ridgway

The largest of New World crossbills, with the deepest and most powerful bill, the depth always 12 mm. or more; in general size averaging appreciably larger than *benti* or *grinnelli;* coloration of adult male deep scarlet to blood red, consequently averaging darker than *grinnelli* with a darker mantle, female averaging darker than *grinnelli,* often but by no means always with a darker, more olive yellow wash below.

Breeds.—Southern Arizona into the mountains of Mexico.

Loxia curvirostra mesamericana Griscom

Much smaller than *stricklandi;* in general the wing measurements are almost exactly as in *bendirei*, but the bill is proportionately deeper and more powerful; in coloration the darkest of New World crossbills, the mantle dark sooty rather than brownish; flanks, vent, and under tail coverts dark brownish-gray; adult males dark scarlet or even dull crimson, appearing even darker than *stricklandi*, but not the dark, dull brick red of *pusilla;* wing of males 89-93.5 mm., average 91.5; culmen 19.2-20.5.

Breeds.—Probably mountain-pine forests of southern Honduras, and north central Nicaragua.

Colorado Crossbills

Five of the eight races recognized by Griscom have been recorded from Colorado. So far as it has been determined, only one race has been known to breed. The remaining four subspecies have been known only from non-breeding specimens. The appearance in

Colorado of these four has been attributed to the crossbill's wanderlust, but it is entirely possible that one or more may have nested in the past, and will be recorded breeding in the years to come. Ranging from their normal breeding areas for thousands of miles, the various forms often invade new territories becoming exceedingly numerous for one or more years, following which they withdraw for indefinite periods.

In explanation, Griscom theorized (p. 82) that two factors might account for their erratic wandering, "although these factors cannot possibly be urged to account for *all* the vagaries of this remarkable bird," namely (1) their dependence upon an ample cone crop which would cause them to shift from one region to another, and (2) competition with the white-winged crossbill.

The former factor is the only one plausible in so far as the races occurring in Colorado have been concerned, since there has been no record published of a white-wing invasion in this part of the Rocky Mountain area.

Comparative data on measurement and color of males for the five Colorado races has been set forth by Griscom (p. 139) as follows:

TABLE NO. 1.

Measurements and color of Colorado races of *Loxia curvirostra.*

Subspecies	Wing	Culmen	Depth of Bill	Color
neogaea	86.5-91mm	15.5-17.5mm	9-10mm	Dull brick to coral red
minor	81-88.5	13.5-15	8-9	Dull brick red
bendirei	86.5-94	15-18.5	9-10.5	Bright scarlet
benti	91-97	17-19.5	10-10.5	Rosy red
stricklandi	98-104.5	18-21.5	12-13	Deep scarlet to blood red

It will be seen from the above that there is an overlapping of measurements in the various subspecies, which often makes identification difficult. The colors listed are for the "red" phase and for mature males of comparatively fresh plumage. There may be many variations in color due to pigment, season, and molt. Extremes in measurements may also be found within a breeding flock (see discussion under *benti*).

COMMON CROSSBILL *Loxia curvirostra neogaea*

The most common and best known crossbill of eastern North America, this race has occurred in Colorado as a rare vagrant. The only specimen known from within the state is a male in the U. S.

Biological Survey collection, taken at Limon, Lincoln County, on November 28, 1914. The bird was identified and listed by Griscom.

It should be noted that Dorothy E. Snyder (1952) has recently published careful observations of the entire cycle of nesting red crossbills in Massachusetts. The notes probably refer to this race. Her study is of particular interest, since it was conducted half the continent away at the same time as the Museum's 1952 study. Also noteworthy is the high degree of agreement found between the two sets of recorded data.

LITTLE CROSSBILL *Loxia curvirostra minor*

Though normally breeding in the Pacific northwest, this form has been noted for its occasional far-reaching invasions. Griscom (p. 121) states:

> in the winter of 1887-1888 a great irruption eastward took place, paralleling the famous flight of evening grosbeaks in 1890. This Crossbill reached the Atlantic States in numbers from Massachusetts to South Carolina, and in the interior south to Louisiana. There is abundant evidence of the return flight northwestward the following spring. There is evidence to suppose that some of these Crossbills have remained in the Boreal zone east of the Rockies ever since, and I have definite breeding records. While there are some records east of the Rockies prior to 1887, there are numerous records for the interior of the continent since that year.

Three of these wanderers were collected near Breckenridge, Summit Co., Colorado by Edwin Carter. A male (D.M. no. 1632) was secured in February, 1878 (day not given); another male (no. 1648) on February 18, 1878; and a female (no. 1649) on March 20, 1878.

Inasmuch as *bendirei* were also secured by Carter in the same general location that year, it is probable that an invasion of crossbills from the northwest occurred, due possibly to a favorable cone crop.

TABLE NO. 2

Measurements of Colorado specimens of *Loxia c. minor.*

	Wing (chord)	Culmen (exp.)	Bill (depth)
1632 ♂ Breckenridge, Summit Co., Feb. ...., 1878	85.7	15.4	8.1
1648 ♂ Breckenridge, Summit Co., Feb. 18, 1878	88.0	14.3	8.1
1649 ♀ Breckenridge, Summit Co., Mar. 20, 1878	80.8	15.2	8.2

BENDIRE'S CROSSBILL *Loxia curvirostra bendirei*

Carter's catalogue lists a nest with young and one egg of a "red crossbill" taken on Goose Creek, Park Co., Colorado on April

12, 1878. It is of interest that he collected three specimens of *minor* in February and March, and two male *bendirei* on April 19, and 24, 1878. Unfortunately there is no skin in the Museum collection secured on the same date the nest was obtained. Therefore the breeding status of the race involved is open to question. It seems probable, however, that it may have been this form.

Griscom lists twelve specimens of *bendirei* from El Paso Co., one from Jefferson Co., and one from Routt Co.

Ten specimens appear in the Museum which can be definitely classified as *bendirei.*

TABLE NO. 3

Measurements of Colorado specimens of *Loxia c. bendirei.*

Males			Wing (chord)	Culmen (exp.)	Bill (depth)
1645	Breckenridge, Summit Co.,	Apr. 18, 1878	91.1	18.0	......
1647	Breckenridge, Summit Co.,	Apr. 27, 1878	89.0	18.0	9.0
12618	Douglas Co.	Aug. 21, 1888	88.1	16.8	......
12619	Denver, Denver Co.	Dec. 20, 1898	90.2	17.2	9.0
24840	Denver, Denver Co.	Oct. 16, 1910	89.1	17.2	......
1628	Leadville, Lake Co.	Dec. 29, 1910	92.5	17.0	9.2
839	Golden, Jefferson Co.	Jan. 12, 1911	87.1	17.2	9.0
Females					
1650	Leadville, Lake Co.	Dec. 29, 1910	85.9	16.4	8.8
12620	Jefferson Co.	Aug. 10, 1887	89.9	16.2	9.0
6584	Sunset, Boulder Co.	May 9, 1917	90.3	17.5	9.6

Griscom states that while he refers "all breeding birds from western Wyoming to *bendirei,* and all breeding birds from Colorado to *benti,* intergradation is completely established by individual variation in these two states as regards size, though not as regards color." This has been found to be true of Colorado specimens in the Museum collection, and the skins mentioned by Selander (1952) may also intergrade with *benti.*

Noteworthy of the males listed above is the fact that four are fully adult and of the characteristic bright scarlet plumage. The remaining three are post-juvenile birds and are of mixed or multicolored plumage.

STRICKLAND'S CROSSBILL *Loxia curvirostra stricklandi*

These, the largest of New World crossbills, occur in Colorado apparently as vagrants. They have powerful beaks considerably larger than other races. Griscom (p. 139) gives the average measurements of adult males as: wing 98-104.5 mm; culmen 18-21.5 mm; depth of bill 12-13 mm. Museum findings agree with these figures except in depth of bill. Adult males from Colorado and a series from Chihuahua, Mexico (Museum of Vertebrate Zoology, Calif.) measured between 11.0 mm. and 12.0 mm. Care was taken in measuring to follow procedure outlined by Griscom (p. 138).

This race was recorded as nesting in Colorado by Morrision (1889), and Breninger (1894), but apparently the adults were not collected. Consequently, there are no authentic nesting records for this subspecies from the state.

The first specimen to be collected in Colorado was a male taken by Aiken, May 22, 1874 in El Paso Co. (Aiken & Warren, 1914). This bird is now in the Museum series (no. 25983). Four other males appear in the collection (nos. 4294, 4296, 4300, and 7418), the first three being taken by F. C. Lincoln at about 7,500 feet elevation near Boulder. These specimens measure as follows:

TABLE No. 4

Measurements of Colorado Specimens of *Loxia c. stricklandi*.

Males				Wing (chord)	Culmen (exp.)	Bill (depth)
25983	adult	El Paso Co.	May 22, 1874	97.5	22.6	11.3
4294	post-juv.	Boulder Co.	Feb. 20, 1915	95.1	19.1	10.8*
4296	adult	Boulder Co.	Feb. 20, 1915	99.9	......	11.2
4300	adult	Boulder Co.	Feb. 20, 1915	101.0	21.3	......
7418	adult	Aurora, Adams Co.	Nov. 2, 1919	104.1	21.7	12.0
Female						
4297	adult	Boulder Co.	Feb. 20, 1915	96.1	20.4	......

* Question might be raised concerning No. 4294, as standard measurements tend towards *benti*. Measuring the width of the upper mandible below the nostrils, the least variable portion of the beak, this bird (8.1 mm.) indicated *stricklandi*, maximum for *benti* being 7.7 mm. The bird's age might well account for smallness of standard measurements.

BENT'S CROSSBILL *Loxia curvirostra benti*

This race, described by Griscom and named in honor of A. C. Bent, is known as the breeding bird of Colorado. Actually, while other races might breed on occasion, *benti* is the only subspecies that has been positively identified as such. Prior to subspecific separation, ornithologists had called the Colorado birds *stricklandi*. Therefore, for the purposes of this study, all such early references have been included under *benti* unless specimens were taken which prove otherwise. It should be strongly emphasized at this point, that all records of subspecific red crossbill identification should be backed by specimens. Few, if any, naturalists are competent to make such determinations in the field, and many specimens in the hand are difficult to place.

Grafton, North Dakota, actually on the northeastern edge of the vast territory over which this race ranges, is the type locality for *benti*. This bird, a male (no. 68343, Zoological Museum, Univ. of Michigan), differs from Colorado *benti* in several ways. Griscom (p. 130) discusses the problem as follows:

> Colorado birds depart from typical *benti* in two important respects. In the first place they are distinctly larger birds, and in the second place the extreme development of the rosy and pale coloration is comparatively rare. It is entirely lacking in some specimens, about half developed or suggested only in the majority. The question naturally arises, what shall we call these birds? I prefer to call them *benti*, as they are nearer that race than any other, showing an approach to the still larger races south and west.

In discussing specimens of *benti* which have wandered from breeding areas, Griscom notes that:

> The vagrants from Kansas are Colorado birds, and so are the specimens from Oregon, New Mexico and Arizona. On the other hand Nebraska vagrants are typical *benti*, as is the straggler from Michigan. So distinct is typical *benti*, that there is no special difficulty in identifying vagrants as either typical or non-typical, and the geographic points of origin of these vagrants is just what common sense would expect.

There are many problems still to be solved, and Griscom concludes:

> The next and final trouble in a thorough treatment of *benti* is the absence of breeding specimens from critical and extensive areas in the Rocky Mountain region west and south of Colorado. Notably there is not a definitely breeding male from either Utah or Nevada in existence. Common sense alone would suggest that in the numerous mountain ranges of these states there must be one or more breeding populations that would connect *benti* of Colorado still more closely with the California breeding bird. As a systematic ornithologist, however, I predict that they will be found, because I have seen vagrant specimens which exhibit the intermediate characters to be expected. These birds come from California, Nevada and Oregon. Some are extremely rosy in color, with dimensions that exceed even Colorado breeding birds, thus approaching the California and Mexican races; others (Nevada, Charleston Mountains) have the dimensions of the California race, but some rosy coloration in adult males in fresh plumage. I also predict that the latter type will be found breeding further west than the first type. In this paper it is to be understood that birds of the first type are regarded as nearer *benti* and the second type as nearer *grinnelli*.

There is another group of intermediates, known only from non-breeding vagrants (eastern Oregon, southwestern Idaho, etc.), which are readily picked out of collections from the western states. These birds are rosy and pale like typical *benti* in coloration, but have the smaller dimensions of *bendirei*. Presumably they represent the fresh fall plumage of the intergrades from eastern Montana, but only breeding birds are known from this region in worn scarlet plumage.

It seems possible, from our limited experience and from the notes in literature, for small breeding populations of *benti* to be scattered over the great areas of the Colorado mountains during each year. On occasion they will appear in great numbers and breed in certain life zones at certain seasons. Two such years, 1947-1948 and 1952, are discussed in following sections.

During these periods of mass nesting the Museum secured an extensive collection of skins representing birds of all ages and plumage phases. Of these, fifty-three birds that were fully adult and participating in known breeding communities were selected for measurement.

At this point, it should be emphasized that the *benti* type specimen was taken in North Dakota, that birds of that area differ in color noticeably from crossbills of the Colorado Rocky Mountain region, and that it is probable there would be justification in naming the breeding birds of Colorado as a new race. For the purpose of this paper, however, Griscom's use of *benti* has been followed.

TABLE NO. 5

Measurements of adult breeding *Loxia c. benti* from Colorado.

No. of Specimens Examined	Measurement	Minimum	Maximum	Mean
Males				
30	Wing (chord)	90.5	99.3	94.10
27	Culmen (exposed)	17.0	20.3	18.70
15	Depth of Bill	9.6	10.5	10.10
Females				
23	Wing (chord)	88.0	94.9	91.24
22	Culmen (exposed)	17.0	20.0	18.23
15	Depth of Bill	9.3	10.2	9.81

The measurements of the males agree closely with those of Griscom's with slight divergence. The females average smaller than the males, as is the case with all races of crossbills. It should be noted that specimens which did not qualify for certain bill measurements were those damaged in collecting or the mandibles had not been closed properly in preparation.

Colorado Nesting Records

There are few accounts of nesting crossbills in the Colorado literature. Morrison (1887) apparently was the first to publish the breeding activity of birds which he identified as *mexicana* (*strick-landi*). He secured a male on January 15, 1886, near Fort Lewis which measured: wing, 3.80 inches; bill, .73 inches. According to current knowledge this bird would probably be classified as *benti*, being too small for *stricklandi*. Later (1888) Morrison reported upon a set of four eggs collected at Fort Lewis, La Plata Co. on January 30, 1887, as "the second set taken in the United States" for the red crossbill.

Breninger (1894) working in El Paso Co. was the next to record crossbills nesting. He collected a young bird January 5, 1893, "evidently a day or two from the nest," and a week later saw a female building. In ten minutes time, the bird "returned to the ground three times, carrying each time a beakful of withered grass to her nest. The male, who is always close to his industrious mate, was perched on a convenient limb, uttering a low sweet warble." Later the nest was found deserted.

Breninger collected six sets of three eggs and two of four during the next four months, illustrating his notes with a photograph of a female upon the rim of her nest, the three eggs showing clearly. This is probably the first photograph of nesting New World red crossbills. He gives (p. 101) the following data:

> All nests were built in positions to receive the warmth of the sun from the south and sheltered from the winds of the north. The distance from the ground to the nest in most cases was thirty feet; two were but eighteen feet from the ground. The nest is deep, with thick walls and warmly lined. The female, who attends to the duties of incubation, sits for days at a time without food or water, depending upon the male for food; which I am led to believe he sometimes forgets, as is evidenced by my examinations of birds shot from the nest during extreme cold weather. In one instance only have I seen the male supplying the setting female with food.
>
> Nests with eggs were found during the months of January, February, March and April, after which no birds were seen nesting or carrying material for nest building. Young birds became numerous, congregating in flocks to feed on pine seeds.
>
> Young birds do not possess the crossing of the upper and lower mandible so peculiar in the adult birds as to be often alluded to as deformities, but appears in stages as age advances. The bird comes into the world with a cone-shaped bill much the same in form as the Crimson Finch (*Carpodacus*

purpureaus), and the full crossing is not attained until the bird is at least six months old.

Henderson (1907) brought to light the notes of Denis Gale, who saw a family of young nearly full grown in the mountains of Boulder Co. on May 21, 1890. Three years later, Gale found a nest, March 31, 1893, on an eastern hillside of Buckhorn Mountain in a pine about eighteen feet from the ground. The female was on two eggs and refused to move until he touched her. Young and old birds were observed on May 8 of the same year. In conclusion Henderson states that, although Gale was an ardent field man, these were the only notes on breeding crossbills in the 278 pages covering the years 1883-1893. He then adds, "although I spend a little time in the coniferous forests of our mountains up to timberline each year, I have never seen any crossbills."

Noting that both Breninger and Gale made their observations in the early part of 1893 at separated points along the front range of the Rockies, it seems likely a general nesting occurred throughout the state. Even more noteworthy, is the fact that for the following forty-three years no further nestings were recorded, even though crossbills were frequently reported and collected.

During the early 1920's the Museum conducted an intensive study of the nesting birds of the mountain area, but no nesting crossbills were observed. Although we have been interested in the species for many years, and have covered many miles through the coniferous forests, we never observed any indications of nesting communities in the 1920's and 1930's.

On April 8, 1936, F. C. Kleinschnitz found one nest two miles south of Parker, Douglas Co., in a yellow pine. Unfortunately, however, it was abandoned shortly after. Apparently there was no community nesting on this occasion, for we made many visits to the area and saw few crossbills. In fact, breeding crossbills were not reported for the next eleven years, until the mass nestings during the winter of 1947-1948.

1947-1948 Season

Coincident with an excellent cone crop in the yellow pines of the Transition Zone, the crossbills appeared on the eastern slope of the Rockies in the greatest numbers as yet on record. They were exceedingly numerous throughout the foothills, and nesting seems to have extended over a ten month period in various life zones. Populations were seen from Wyoming to New Mexico and, according to specimens collected, they were all of the race *benti*.

The first breeding observations came from Snyder and Cassel (1951) in late July.

On July 26, 1947 while censusing the breeding birds of a forest of lodgepole pine (*Pinus contorta*) near Science Lodge, University of Colorado Mountain laboratory (9500 feet elevation), we saw a female Red Crossbill carrying a piece of nesting material to her partly completed nest 18 feet up in a 20-foot lodgepole pine. When we returned to the nest on July 29, it was completed and contained one egg, but there was no sign of either the male or the female. Apparently incubation did not begin with the laying of the first egg. In a spring nest in Ontario, Ross and Ross (1950) found evidence that incubation did begin with the first egg. On the afternoon of July 30, however, the female was observed on the nest at 4:30 P.M. and again at 7:00 P.M. Incubation probably had begun. We did not ascertain how many eggs were in the nest for fear of alarming the female. During the next few days we visited the nest several times, finding the female incubating each time.

On August 7, the authors made careful observations throughout the day, from 4:30 A.M. until dusk at 7:30 P.M.

From 4:30 to 7:00 a.m. the female stirred only occasionally and did not defecate, preen, or stretch. At 7:06 the male came with food. The female chirped loudly, lifted her head, and opened her mouth. After feeding his mate the male flew to a pine close by, where he chirped briefly. At 8:16 the female departed, perhaps responding to the call of her mate (we heard a crossbill in the distance). In about three minutes both returned to the nest tree. The female quickly went to the nest where a few seconds later, the male fed her. The pair then left again, flying off together. After three minutes the female returned to the top of a nearby tree and then quickly went to the nest. Presently she exchanged calls with a crossbill we did not see. At 9:09 the female climbed out of the nest, perched briefly on a branch close by, then went back to the nest. At 9:57 she again exchanged calls with a bird we did not see. At 10:07 the male appeared, perched 25 feet away for a moment, then flew to within three feet. The pair then left the nest, flying in different directions. At 10:09 both returned, the female to the nest, the male first to the top of the nest tree, then to a nearby tree. After several minutes of exchanging chatter with his mate, the male flew away. At 12:16 the female called and the male returned, chirping, fed the female, and flew off. At 3:18 the male reappeared calling. The female answered and left the nest, followed shortly by the male. At 3:23 both birds returned, the female directly to the nest, the male chattering, to a tree 25 feet away. At 3:25 he ceased calling, fed for about fifteen minutes, and then left the immediate area. At 4:51 he returned, calling, and alighted in a treetop 20 feet from the nest. The female answered and left the nest. The male then flew off, again in a direction different from hers. At 5:04 both birds returned, the female flying from tree to tree before going to the nest, the male perching in a treetop and chattering quietly. At 5:12 the male flew off. At 5:40, coincident with the falling of a little rain, the female exchanged chatter with a crossbill we did not see. At 5:45 a few more drops fell, and again the chattering commenced. The rain and wind continued intermittently until dark (7:30), during which period the female remained on the nest, and we saw no sign of the male.

Photograph by Alfred M. Bailey & Robert J. Niedra
February, 19

BENT'S CROSSBILL

The high plumaged male posed on the nest's rim.

Photograph by Alfred M. Bailey & Robert J. Niedrach
February, 1948

BENT'S CROSSBILL

The female was besieged by the twelve-day-old young.

YELLOW PINE COUNTRY

Yellow pine and shrub oak association showing the 1948 nesting area of crossbills in Douglas County.

Not only were these observations the first all-day record of an incubating female, but the first of a Colorado nest in the summer, in the Canadian Zone, and in a tree other than yellow pine. This was the first clue to the seasonal shift of breeding *benti* through the life zones.

Two months later on September 9, a second nest was reported to the Museum by Joseph C. Howell. Located in a yellow pine, it was found in Moraine Park of Rocky Mountain National Park. The female proved quite tame, and when disturbed "scolded with excited chirps."

Throughout the fall months, the Museum staff saw many crossbills, adults and young, in the pines south of Parker, Douglas Co., and a fine series of specimens was taken. Unfortunately a search for nests was not undertaken until late January, 1948. Neidrach found the first, with three eggs, February 1, on the south side of a yellow pine. During the next two months, he located eighteen additional nests in the same general area, either with eggs or young. Without exception all were on the south or southeast sides of the trees.

The breeding cycle continued through April, the height apparently being reached in the early part of March. The majority of the nests were inconvenient to photograph without carrying heavy ladders into the hills, so no effort was made to obtain pictures until April. Several nests were found during the first two weeks, and we secured photographs of three near Daniels Park.

The last nest of the season was found on April 18 in the yellow pine forest south of Franktown in Douglas County. Pasque flowers were in bloom on warm slopes and bands of crossbills were moving through the dark forests. Small groups were feeding on the ground in the shade, and on occasion they would be joined by pairs evidently nesting. By searching the south sides of the pines, Bailey located this nest ten feet from the ground and far out on the tip of a branch. This, the lowest *benti* nest on record, was well concealed by the long leaves of the pine. Climbing above the nest site, he shoved the protecting vegetation aside with an outthrust foot revealing the gray-green female fearlessly incubating. When we returned to the nest two days later, April 20, the eggs had hatched and there were three small young. Before we could get a ladder in place the adults arrived and fed the nestlings. They worked at double speed, keeping a wary eye on the would-be photographers while chucking food down the gaping uncrossed beaks. Both adults then departed; the female returned in ten minutes, fed the young, and then settled down to brood. After about twenty minutes the brightly colored male alighted on the top of an adjacent tree, which proved to be his favorite song

Photograph by Alfred M. Baile
April 24, 194

CROSSBILL COUNTRY

A heavy, wet snow blanketed the nest site (X).

Photograph by Alfred M. Bailey
April 24, 1948

BENT'S CROSSBILL

The female returned to feed her six-day-old young after the snow was cleared from the nest.

perch, and then flew down to the nest. He slowly regurgitated a portion of his cargo to his mate, then both stood on the rim of the nest and fed the feeble babies.

April 21 was an overcast day and at 6 A.M. there were a number of crossbills in the vicinity of the nest, mostly adults with young. There was much more activity in the yellow pines than we had seen previously. Crossbills were everywhere; bands of adults and young fed upon the ground, and there were numerous other species working the near vicinity. The adult crossbills came to their nests at about twenty minute intervals during the first two hours, and on each occasion the male shared a portion of his food with the female, and then both fed the young. The following day we reached the nest at 5:45 A.M. and the birds returned three times while we were there—at 6:15, 7:25 and 8:45. They were in no hurry, both perching high in the adjacent tree, then leisurely flying to the nest to feed their young. The female did not brood as she had done on previous mornings.

There was a severe snowstorm on April 24, and on the following day Bailey headed for the nest site alone. Leaving Denver at 6:00 A.M., he hoped that by some miracle the nest had been able to survive the storm. His notes for the day read:

April 25, 1948. Had a heavy, wet snow yesterday which blankets all of eastern Colorado. More than 1000 people were marooned along the Colorado Springs Highway last night.

I started for the crossbill nest at six, hoping the young crossbills had been able to pull through the storm. The graders had roughed off much of the snow from the road, but there were numerous cars stalled along the way. The sun came up clear with a promise of a beautiful day, but the nearer I came to Franktown, the heavier the snow. Finally came to the county road and the grader had been through once, but I could not make any headway. Got turned sidewise and had to shovel out. Just as I was contemplating starting back to Denver, the grader came along and made a swath to the top of the hill, and I followed behind.

Started to walk through the knee-deep snow through the pines. It was a fairyland with all the trees massed with white, but the going was too heavy, so I left the ladder and broke trail to the nesting tree. As I came over the rise, I saw the male crossbill on top of the pine about fifty yards from the nest site. It was his favorite perch, but one glance at the nesting tree showed that only a miracle could have saved the nest. All the branches were massed with snow, and bent down so I could not tell which limb had supported the nest. I dug around in the snow on the ground to see if I could find any of the babies from the nest. And just as I was ready to start home, the male fluttered down, looked me over, and then flew under a bent branch, and disappeared from sight. He had a little tunnel through the snow, and there was the nest at a 45 degree angle with the young clinging to their precarious perch. They could not fall out, for they were completely covered with snow.

Photograph by Robert J. Niedrach
May 1, 1948

BENT'S CROSSBILL

When eleven days old, the nestlings forced the male to stand on tip-toes while he fed them.

The day was warm and the snow started falling from the trees in great masses. Fortunately the nest was not struck.

Returned to the car and packed the camera in, set up on the nest site, and while the camera ran, I knocked the remaining snow from the nest. Then returned, dragged in the ladder and set up. The female covered the cold wet young. A few pictures were shot.

The next visit was made on May 1, when the young were eleven days of age. The old ones were more deliberate about returning to feed the young; in fact, they did not show up for an hour and a half. The young were so large they almost concealed the adult male on the back rim of the nest. The young extended their necks so high that he almost had to stand up on tiptoes to shovel food into the wide open beaks of the youngsters—now well covered with their gray juvenile dress. The male and female young are colored alike in their first plumage, and the beaks of the youngsters are short, and not crossed.

No other nests were found during 1948, and only a few crossbills were observed in the area adjacent to Parker during the next few years. In January of 1952, Don Thatcher located a small community of crossbills in the pine clad foothills west of Denver, and he and Lang Baily found the first nest in the area on January 16, near Hosa Lodge in Jefferson County. The nest was on the south side of a yellow pine about twenty-five feet from the ground, and we took a ladder to the nesting tree on January 18, and made a series of photographs of the adults returning to their three-day old young. The nest was about eight feet from the trunk of the tree, and three feet from the tip of the branch; so we placed a strobe light four feet from the young, and anchored our 35 mm. Exakta with the 240 mm. lens in a strategic position. The actions of the birds were similar to those we had photographed at Franktown and Daniels Park in 1948. Both birds returned—a little hesitatingly at first, naturally, for we were both on the top of the ladder in plain sight, and only eight feet away. The male perched on the top of a nearby dead pine while the female remained a few feet from the nest summoning her courage, but within fifteen minutes she was with her young, feeding them by regurgitation. Almost immediately she was joined by her mate, an exceptionally brightly colored individual.

The old ones fed deliberately, giving their young quantities of light colored food—probably pine seed pulp—until the neck membranes of the little fellows were extended into large bumps. When the adults left, they flew off together, but Baily noted that they separated shortly, flying different directions. When they returned, however, they arrived at the same time. They came in at approximately forty-five minute intervals. On the second visit after our arrival, the female crouched in the nest rim with quivering wings and open beak, begging the male for a portion of his cargo, but he ignored her and chucked food down the throats of the young. We took numerous photos, the flashing of the strobe light just four feet away apparently not being noticed by the birds.

Lang Baily discovered another nest on January 26, in the crown of a small yellow pine, about twenty-five feet from the ground. The set of three eggs was completed on February 6, and we visited the nest the following day with him. A ladder was put in place, the top being just two feet from the nest; the camera and strobe light was lashed in place. Both of the adults sat in a neighboring tree showing little concern. We retreated fifty feet away to watch, the female immediately returned to the nest and settled upon her eggs.

It was a warm day well above freezing, but we waited five minutes to allow the female to take the chill from her eggs, and then Bailey slowly climbed the ladder, talking as he ascended so the bird would not be startled by his appearance. She sat tight, facing the strobe, camera and photographer just two feet away—and did not blink when the light flashed at 1/1000 of a second. Numerous pictures were made showing the eggs and the adult coming to her nest. Later, after a snowstorm massed the yellow pines of the area, Lang Baily secured pictures of the female upon her snow rimmed nest.

Baily and Thatcher had located many nesting pairs in the near vicinity. Although no concentration of large numbers of birds was observed as in 1948, it was evident that many pairs of crossbills were active, and that it might be an excellent opportunity to make a worthwhile nesting study of these unusual birds. Consequently, Lang Baily was delegated to make the survey and his report to the Museum on the nesting community is presented as Part II of these notes on the red crossbills of Colorado.

Since this study was completed, a year of field observations on Colorado crossbills has been accumulated. Crossbills were found as common throughout the Hudsonian and Canadian Zones during the fall and winter of 1952-1953. Only a few birds were reported from the Transition Zone. Lang Baily observed the birds gathering grit in a shelter house at Echo Lake throughout the winter and spring of 1953. Though he suspected breeding, young birds were not seen until April 26. During June and July roving bands of birds in postjuvenile molt were constantly seen in the area indicating that breeding had been more extensive than previously observed. Then, on July 8, 1953, Baily found a pair building on the Genesee tract coincident with the promise of excellent cone crop in the yellow-pines of the Transition Zone. This observation completed the record for crossbills nesting in every month of the year in Colorado.

BENT'S CROSSBILL

The female feeds the young which were approximately three days old
(Nest 1)

Photograph by the authors
January 18, 1952

BENT'S CROSSBILL

The female begs food from the male, and then both feed

Photograph by the a
January 18,

Photograph by the authors
February 7, 1952

BENT'S CROSSBILL

The female returning to incubate.

Photograph by the authors
February 7, 1952

BENT'S CROSSBILL

"She sat tight" eyeing the photographer (Nest 2).

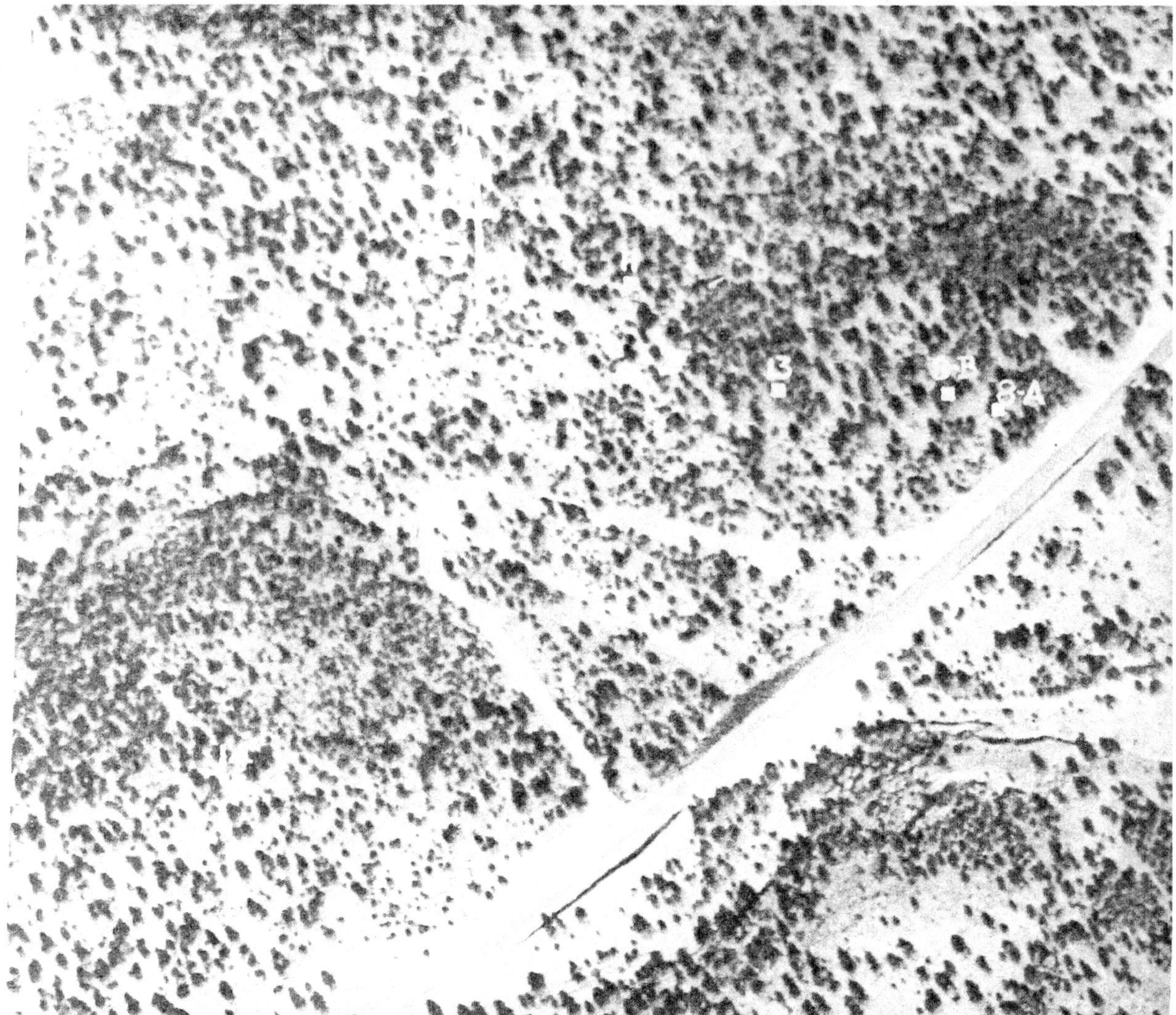

TRANSITION ZONE

Aerial photograph showing 1952 nesting sites
(Nests Nos. 3 and 5 extralimital).

Part Two

1952 Observations of the Red Crossbills

By A. Lang Baily

Early in January, 1952, Don Thatcher reported regular singing activity of red crossbills, *Loxia curvirostra benti*, on his winter census tract near Genesee Mountain in Jefferson County, fifteen miles west of Denver. On January 16, a cold and windy day, Thatcher and Baily located a nest containing three young crossbills approximately four days old. This initiated the following general study of crossbills, which continued to breed in that area until late June, and in the high mountains until September. A total of 250 hours were spent in the field, involving the co-operation of Dr. Alfred M. Bailey, Les Baylor, Reginald Denham, Robert P. Fox, Ferd C. Kleinschnitz, James Muir, Robert J. Niedrach, Don Thatcher, and Harriet B. Woolfenden.

Of an estimated thirty pairs nesting locally near Genesee Mountain nests of fifteen pairs were located within an area of a square mile formed by a watershed draining to the southeast. The irregular topography was covered by yellow pine (*Pinus ponderosa scopulorum*) associated with Douglas fir (*Pseudotsuga taxifolia*) on northerly slopes. Typical of the yellow pines, open patches of ground were common between groupings of trees, thereby supporting many herbs and shrubs of the Transition Life Zone. Temperatures (F.) ranged from a low of -10° to a high of 45° during the winter months and rose to a low of 55° and a high of 85° by the month of June. Light precipitation fell in January and February, while heavy snows were common through March and April, followed by afternoon showers during May. Little, if any, rain fell for the greater part of June.

When found, each nest and respective pair of birds was assigned a number. All subsequent observations were listed accordingly. During each trip to the area every active nest was visited. Unfortunately, it was not possible to make day by day observations and accordingly, much information remained ungarnered.

Banding was attempted on a small scale. Due to lack of success in luring adults to traps, the only mature birds banded were females which could be taken off their nests. Twelve young birds were ringed just before fledging, and weights and measurements were secured with each banding. It became quite evident during the study that extensive banding might proved conclusively many important facts concerning their habits. Until this can be done, however, the following digest of the Museum's 1952 observations is being offered to those interested in the red crossbill.

Flock Behavior

In order best to comprehend these intriguing nomads, they should be observed in the field. As previously mentioned crossbills are known for their wanderings and gregarious habits. When in a *transitory* or nomadic flock, they often feed in the same tree, chattering softly in accompaniment to the sharp cracking of pine seed. The branches appear alive with movement, the birds bending and twisting over terminal clusters of cones, accented by the audible flutter of wings as fresh cones are investigated and better footholds gained. Then, on sudden impulse, the entire flock will rise and wing swiftly to a neighboring tree or distant hillside, the quiet chatter changing to loud, clear calling. Where food is plentiful, many such flocks may be found on high mountain slopes, the air filled constantly with their voices. On the following day, however, the forests might be quiet and deserted.

Sometime during the sequence of seasons, the exact stimulus being unknown, they may initiate a breeding cycle. Even then, their group instincts are not abandoned. Nesting in extensive communities covering several square miles, they associate closely when not occupied with activities at the nest. Such *foraging flocks,* of loose knit structure, have been observed regularly with breeding Colorado crossbills. Composed of unemployed birds, males feeding incubating females, pairs feeding nestlings, and birds with fledglings in tow, these flocks are in constant fluctuation, with arrivals and departures to and from the nests. On another day, however, the foraging flock may not form, the birds working in isolated pairs or family groups.

Foraging flocks frequently are found with typical winter association groups composed of woodpeckers, nuthatches, chickadees, purple finches, siskins, and juncos. There, the crossbills play the role of predator lookout, since they usually feed in the upper portions of the trees. It has been noted, however, that individuals maintain a degree of independence, lagging behind or diverging from the main flock when it pleases them to do so.

Plumage

Due to the irregular habits of the crossbills, it has been difficult for ornithologists to obtain accurate information on their plumage sequence and coloration. There is a wealth of material in the literature, but there is, however, little agreement. This is understandable, since short term observations record only a particular instance that may be affected by many governing factors.

Two basic colors appear in the male plumage, red and yellow. This dichromatism occurs in other related finches, but it is not so pronounced or puzzling as in the crossbills. The red plumage is thought by many as typical of the adult, yet individuals can be found with yellow pigmentation only, others with a mixture of both, resulting in a *multicolored* or *xanthochroistic* plumage. In fact "red" males will often have a few yellow feathers scattered over their bodies. Many theories explaining this variation have been advanced based on age, diet, hormones, climate, season, and pigment change within the individual feather. Laboratory experiment is difficult, since captive crossbills usually acquire the yellow or multicolored dress. Future field observations, examination of skins, and controlled experimentation will, no doubt, provide a synthesis of the many theories. It has been in this light that following discussion of the Museum's observations has been written.

During the breeding cycle near Genesee Mountain the majority of males were of the typical "dull-red" plumage for Colorado *benti*. Two males were observed, however, wearing the multicolored attire. All of these birds were apparently fully adult as evidenced by specimens collected. Yellow males were not seen, nor have any ever been collected in the state as far as has been ascertained.

As summer arrived breeding continued at higher elevations. There flocks of juvenile birds became evident by late June. Averaging from twelve to fifty in numbers, these flocks were composed of birds, male and female, either in streaked juvenile plumage or in the process of postjuvenile molt. Noteworthy was the fact that *all* males in postjuvenile molt were acquiring the multicolored attire. By October all breeding had ceased and the majority of the population had disappeared from the high country. Those males that remained were once again typical red with a low percentage of multicolored individuals.

Of particular interest was an exception, a breeding adult male collected in mid-July of *particolored* plumage. Certain areas of the body were all red, while others were all yellow. This bird had just started to molt on the wing, the body feathers being old and worn resulting in unusually bright pigmentation.

If conclusions are to be based on these and previous Colorado observations, as well as the specimens in the Museum collection, it would appear that the majority of red males were adult, the majority of multicolored males were postjuveniles, with a scattering of red

BENT'S CROSSBILL

Photograph by the au
January 18,

postjuveniles and multicolored adults. Particolored and yellow males would seemingly be a rarity. However, despite this consistency of the Museum's findings, there is considerable evidence to the contrary concerning races outside of Colorado. In fact Tordoff (1952) considered the majority of postjuvenile *benti* wintering in Kansas to be red. Majority of postjuvenile *benti* wintering in Kansas have red plumage.

Jollie (1953) offers a solution to this problem. He advances the theory that there is second postjuvenile "red" molt following the multicolored plumage. This might well account for Tordoff's wintering red postjuvenile and the Museums multicolored postjuvenile following a breeding season.

The female crossbill has been found to be much more consistent in coloration. The pale yellow-green and gray plumage varies only slightly within a race. A few exceptions have been noted in the literature. Tandy (1889) discusses the assumption of the male plumage by the female.

Museum observations on the time of annual molt agreed well with Ticehurst (1915). Both juveniles and adults apparently molt during the summer and early fall. Some individuals start early, others late, the height of molt being July through August. A series of skins from the Museum of Vertebrate Zoology demonstrated that the western race *grinnelli* follows the identical pattern.

Of interest was the fact that the annual molt can occur irrespective of breeding activity. This was demonstrated by a breeding male (D.M.N.H. No. 26613) secured in July 1952.

As observed from collected specimens the primaries start to molt with the ninth, progressing outwards towards the first. The body molt was found to start on the head, throat, and breast after the seventh primary had fallen. In the case of postjuvenile molt, however, feathers of the tail, wings, and portions of the wing coverts are retained. This fact proved an invaluable aid in determining age. The buffy tips of juvenile wing coverts and the lack of red or yellow pigmentation in the outer webs of the tail feathers marks an apparently adult specimen as a postjuvenile bird.

Feeding Habits

The mandibles, from which the name "crossbill" is derived, is a phenomena of considerable scientific interest. Miller (1914) found that only the horny sheaths are affected by the crossing, the bony foundation of the mandibles remaining untwisted. The actual crossing process occurs shortly after the young are fledged. Prior to that

Photograph by the autho
February 17, 19

BENT'S CROSSBILL

Incubating females prove fearless, often accepting food from the hand (Nest 6).

time their beaks are normal and finch-like in appearance. Of interest is the fact that there seems to be no particular rule governing which direction the mandibles will cross, to the left or right. Miller found an equal proportion of both irrespective of sex. Among 77 males examined in the Museum collection, 45 of the lower mandibles turned to the left, as opposed to 32 to the right. The females balanced more evenly with 24 crossing to the left and 23 to the right. The lower mandible served as criterion of crossing, since it consistently has greater torsion than the upper mandible.

The use of this specialized beak has long been subject to controversy. Robbins (1932) fortunately cleared the air through his observations of a captive bird feeding on *Pinus rigida*. Following the sidewise insertion of the mandibles between the cone scales and their rotation towards the vertical, he states:

> Further, as the rotary movement progresses the upper mandible would ultimately be forced into position so as to act much like a lever,—its tip resting on the lower scale, the broad curve above its tip against the underside of the upper scale . . . the lower mandible automatically relieved of pressure, can open (while the scales are being held apart) and thus give free action to the prehensile tongue.

1952 findings were in agreement with this description, though the method of observation varied. Using an incubating female, hemp seeds were proferred, held securely between thumb and forefinger, thus simulating the scales of a cone. In this manner, the observer not only was able to see but also to feel the deft way in which she flicked the morsel into her beak. The seed, constantly controlled by the tongue, was then quickly husked between the adjoining lateral edges of the mandibles.

This ability of the female to accept and utilize food undoubtedly foreign to her experience, furthered the theory that crossbills can handle many types of food with ease, despite their specialized beaks. Such adaptibility indeed would be a necessity for survival of these birds during seasons of coniferous seed failure. Much evidence in support of this belief has appeared in the literature. Besides being observed eating the seed of almost every North American conifer, crossbills have been recorded as eating almonds, sunflower seeds, hemp, the buds of cottonwood, gum, spruce, frozen apples and pears, as well as succulent parts of eucalyptus. They have been known to take a variety of insect life, including caterpillars, aphids, and several types of tree galls, their crossed beaks proving useful in the manipulation of the latter.

Besides organic material, crossbills have been observed taking several inorganic substances other than grit. Salts have appeared to be a necessity for their diet as recorded by Aldrich (1939), Fisher (1888), Marshall (1940), Meade (1942), and others. Also, during the breeding cycle at Genesee, Baily and Thatcher observed crossbills securing scrapings from the bleached bones of deer.

Song

A knowledge of the accompanying vocal expressions has been found as fundamental to the interpretation of the crossbill's behavior. It was discovered that an experienced ornithologist, without seeing the bird, can tell its activity of the moment merely by hearing the song or note. The following discussion, therefore, should be supplemented with the reader's own field observations, if such insight is to be gained . . . The written descriptions are to be considered at best as mere clues.

Libidinal Song. This was found to be the most common song, given only by the males. Though sung throughout the year, it was best developed during the breeding cycle. In essence, the phrasing was simple and finch-like, resembling in cadence the opening bars of that of the song sparrow, best verbalized by Silloway (1905) as *preet-preet-preet . . . oree-oree-oree.*

While breeding, the song usually was presented from a conspicuous perch located within fifty yards of either the nest or the female. It also was found that the lower the singing perch, the softer the song, and the closer proximity of the female. This rule was an invaluable aid in locating nests. Only once, during the course of the study, was this song given loudly at the nest site itself and then prior to the start of construction.

Among non-breeding flocks, the libidinal song was used widely by males of all plumages. Song development, however, was found to be related directly to both the age and sexual potential of the male, as demonstrated by a series of transitory breeding males collected while singing near Mount Evans in July of 1952. Among juvenile birds this song was often reduced to a series of rough squeaks.

Nuptial Song. Coupled with nuptial flight, this song was the most spectacular given by the male crossbill. Starting with the libidinal song, the bird launched into a continuous series of phrases, running the gamut of finch-like inflection and intonation. Though associated for the most part with flight, it occasionally was presented from a stationary perch. The song usually occurred early in the breeding cycle, prior to the laying of the eggs. Here again, the locale of presentation was determined by the location of the female or the nest site.

Traveling Call. Used by both sexes, this call consisted of three connected notes given in repetition: *quit-quit-quit- . . . quit-quit-quit*, etc. Its projected delivery almost always was heard when crossbills were in flight. If given by a single bird or a pair passing overhead, it often was a good indication that nesting was in progress. The call also was observed as frequently given prior to flight at the termination of feeding or similar activity.

Communication Call. A variation of the traveling call, this call usually was uttered quite softly, depending upon the proximity of the birds to each other. The three note sequence was not strictly followed, a single *quit* being common. Both sexes used it constantly in all social activities other than flight.

Alarm Call. This was a loud, rapid and projected delivery of the traveling call, the intensity of which signified the degree of concern or danger. At the nest site, however, the birds would remain silent until discovery by the predator was a certainty.

Pre-Nesting Behavior

It has been felt than many questions remain to be answered concerning this period of the crossbill breeding cycle. The initial stimulus for the cycle, prevalently theorized as being related to a heavy coniferous seed supply, has yet to be scientifically proven. A breeding male, collected at 11,000 feet in July of 1952, was found to be feeding entirely on aphis (*Dilachnus palmerae*), an exception well worth noting.

The process of pair formation also remains to be satisfactorily determined. At no time during the study conducted near Genesee Mountain were we aware of clearly defined mate selection. Instances of chasing between males within foraging flocks were observed on occasion, but interpretation of such actions was felt to be unwarrantable. The following behavior patterns, however, were observed.

Nuptial Flight. Commencing from a tree top, the male would describe an arc or circle above the forest, singing the nuptial song previously described. A quick three-quarter stroke was characteristic of the wing beat. At times, this display was modified to a short flight between trees, an upward or downward sweep of direction being common.

On January 18th, seven days prior to the start of construction, the nuptial flight was observed by R. J. Niedrach at nest site No. 2. This was the only instance among many that the flight was related to the nesting of a particular pair of birds. At Genesee Mountain, such activity was common in January, decreasing thereafter until the middle of February, at which time, it ceased entirely, nesting activity having been well started throughout the area.

Nuptial Feeding. This was first observed by Lang Baily on February 8th, while working the area south of U. S. Highway 40. His notes were as follows:

> Two pairs were together in a juniper (*J. scopulorum*) surrounded by young pines. They were quite active, chasing up and down through the branches, often alighting on the ground. At times, a female would let herself be "caught" and fed by a male, during which she would droop her wings with small vibrations and chirp much like a fledgling. This was repeated several times before they departed across the valley.

Nuptial feeding was observed next during the construction of nest No. 8a on February 20th. The birds had paused from their building activity to feed upon fallen pine seed. Terminating their foraging, the male fed the female, then both birds departed downslope in search of more nesting material. This was seen again by Baily on March 16th, immediately following the selection of a nesting site by pair No. 13. A fourth instance was recorded when a pair of birds terminated their grit collecting at a picnic site. In this case, however, no connection with a nest was established.

These observations indicated the relative importance of nuptial feeding in the breeding cycle of the crossbill. Apparently it is consistently performed throughout the pre-egg period, ceremoniously accenting daily activities. Later it blends with the male's feeding routine during incubation. Hoffman (1926) offers an additional observation of nuptial feeding when, following the observed act, he collected the female and found that she contained an egg ready to be laid.

The chronological place of copulation in relation to the breeding cycle as yet has remained unknown. Only one instance of mating was observed in 1952 and then by an unidentified pair of birds.

Territorialism

With many passerine birds, a strict observance of nesting and feeding territories has been found. The crossbills, due perhaps to their gregariousness, showed small interest in such matters. During the study, there was but one instance of territorial dispute observed.

On March 16th, Baily was attracted by a red male singing lustily his libidinal song on a ridge north of the highway. Shortly thereafter, the bird flew fifty yards to the northwest, where he repeated his song from the top of a tall pine. Immediately, the bird was challenged by a multicolored male whose nest was 200 feet distant. Perched in the same tree, a vocal argument developed, climaxed by a descending aerial struggle. On striking the ground, the birds separated, the multicolored male departing towards his nest. The red male then moved to the east, where once again his presence was opposed by an established male. Following a headlong chase through the trees, the air-borne battle was repeated, the contestants separating just above the ground. His east and northwest boundaries evidently determined, the red male returned to where his song had first been heard. There he produced his mate and shortly thereafter, they were observed in the selection of nest site No. 13, as proven by subsequent visits to that particular location.

As the season progressed, it became evident that once a nest had been established, territorial lines were forgotten. In fact, it was not uncommon to see small groups of crossbills "visiting" a nest tree without objections from the resident birds. Niedrach further confirmed this visiting habit from his experiences with nesting crossbills during 1947-1948.

Nest Site Selection

Of the sixteen nests located between January and June, only three were observed during the actual site selection, and in two instances only portions of the procedure were witnessed.

On January 25th, where on the 18th the nuptial flight had been seen, the birds were found moving through the young pines, the male often pausing to sing softly. The female kept out of sight, working the inner portions of the trees. Shortly, both birds disappeared in the top of a 26 foot yellow pine. After a lapse of a minute, the male flew to the terminal tip and sang strongly, the female departing down the hill. Momentarily leaving his perch, the male reappeared and resumed his song with a dead pine needle held in his bill. This was indeed an accomplishment, and the definition of his song suffered considerably. Two minutes later, he placed the needle at his feet and flew off, presumably to join his mate.

Again, on Feb. 28, the day after the first nesting attempt of pair No. 8 had been destroyed, the selection of the second nest site was observed in part. Though the moment of selection was not seen, the behavior of the male was noteworthy. Every few minutes he would fly into a densely needled pine, each time stopping at a particular place on one of the horizontal branches. There he sang and called loudly before whisking out and down ridge to where the female was known to be feeding. This in itself was unusual, since never had we heard a male sing with such energy without being on the most conspicuous perch available. Though the performance was repeated time and time again, the female never came near the spot. On March 5th, however, a nest was well under construction at that very location.

The third instance was at site No. 13, previously mentioned in connection with territorial defense and nuptial feeding. Here the complete process was seen. The male, having defined his territorial boundaries, joined the female and both birds began a silent search of the lower branches in the area. Starting at the outer extremities of the limbs, they worked toward the trunk, pausing here and there to pry at pieces of bark or to rest motionless for a moment or two. Only on occasion a barely audible communication note was detected. Finally the female, which had consistently taken the initiative, stopped on a horizontal fork canopied with dense needles. There the male, on some unheard signal, came and fed her. Whether food was actually transferred remained undetermined, though the negative was suspected since the male had not been observed to feed during the previous hour. Following the ceremony, the pair became vocal, exchanging communication notes, then both departed out over the forest giving the traveling call. A subsequent visit on March 25th, despite a heavy snowfall, found the birds actively building on the sheltered fork.

Even as there was variation in the choosing procedure, there was variation in choice of nest locations. All, however, were in yellow pine, though Douglas fir was often closely associated. Also, the nests never were found in densely timbered areas, but rather in fairly open growths, or semi-isolated groups of trees. Seven of the sixteen nests were within 200 feet of a dead or dying pine, used by the male for singing, preening, and as a cautionary stopping place before entering or leaving the nest area. Of the trees selected, ten were between 42 and 50 inches in girth (chest high), ranging in height from 40 to 54 feet. Among the remaining nest trees, one was 26 feet high, the others extending upwards to 73 feet.

The nests themselves were located from 15 to 48 feet above the ground, the majority between 20 and 40 feet. Ten of these were located on lateral limbs exposed from southeast to south to southwest. Two others were facing north, while only one faced due east. The remaining three nests were situated in terminal growth. It was of interest to note that the latter nestings were all successful in fledging young, while only two of the lateral limb nestings were completed. All nests were never more than five feet from the end of growth and never less than two feet. (See table No. 7).

Each nest was situated so that sunlight would reach the protecting needle cover or the nest itself at some time during the day,

YELLOW PINE ASSOCIATION

A typical nest site (arrow) with southern exposure.

usually in the early afternoon. Rarely could the nest be seen from below, though on one occasion, a bird built on a north limb so poorly protected that the incubating female could be seen from all directions.

Throughout the entire study, the minimum distance between active nests was recorded as 225 feet, as found separating sites Nos. 6 and 10.

Nest Building

It was during the building process that the nests were easiest to find. Both adults were quite conspicuous and could be approached closely as they searched through the ground litter. The female took the initiative in collecting and arranging the majority of the material. During a ninety minute observation of construction at site No. 2, the female brought fourteen loads of material, as compared with the male's two contributions. Yet, despite his lack of a helping hand, the male was in constant attendance on his mate, occupying his time by (1) keeping a lookout and singing above the nest or near by, (2) accompanying the female on her trips, but rarely lending aid, (3) entering the nest site and rearranging the pine needles around it, and (4) making his own excusion for material, choosing hastily, dropping most of it on the return flight, and inserting the remainder before the female's return.

The female, on the other hand, took considerable care and time in the selection of the nest material. Several minutes often elapsed before she found a piece that suited her desires. Three or four selections usually constituted a load, then with a soft communication note to the male, she would fly directly to the nest. There, she spent an equal amount of time arranging the material.

Building the nest was not a continuous process. Absences of one or two hours were common both during the morning and afternoon. These apparent excusions for food delayed construction considerably, extending the building, from start to finish, over a period averaging five days.

Merrill (1898), referring to the collection of nest material by crossbills in Idaho, says, ". . . a constant habit being to do this at a considerable distance from the nest." Contrariwise, we found during our study that most of the material used was secured within a radius of 40 yards of the nest site, each nest reflecting the plants and ground litter of the immediate vicinity. Only a few long distant flights were observed and then for chicken feathers or meadow herbs.

The basic foundations of the nests were consistently similar, usually composed of stems and twigs of golden banner (*Thermopsis divaricarpa*), mountain mahogany (*Cercocarpus montanus*), mountain sage (*Artemisia frigida*), squaw currant (*Ribes cereum*), and Douglas fir (*Pseudotsuga taxifolia*). This often was varied with the addition of local items such as stiff feathers, picnic litter, or stems of other associated plants. Yellow pine needles and twigs were noticeably absent in the construction.

As the building progressed toward the nest bowl, more and more local grasses were added, usually soft disintegrating growth from previous seasons. The final lining consisted of the softest fibers, an occasional feather and, if in local supply, old man's beard lichen (*Usnea barbata*). Binders were used on occasion, consisting of small cocoons and web from the tent of the larval tiger moth (*Halisidota ingens*).

There was but one abortive nesting attempt observed. The construction of nest No. 5 was abandoned after the framing had been completed. There was no further breeding activity in that area following abandonment.

Egg Laying

Four, sometimes five, days elapsed after the completion of the nest before the first egg was laid. During this period, the parents rarely were seen at the nest. In fact, the first observation of this occurrence almost convinced us that the nest had been deserted.

The eggs were laid at the rate of one a day, the time of day as yet to be determined. Sets ranged from two to four in number, three being average. Of 12 nests in which eggs were observed, two were complete with two eggs, seven with three, and three with four. The sets consisting of only two eggs were definitely established as second nestings, the previous attempts having been disrupted by predators. Of particular interest was one of these second attempts since the birds had used the same nest for both sets of eggs.

Considerable variation appeared among the eggs themselves, measuring in millimeters from 22.6 to 21.3 in length and 16.4 to 16.0 in width. At seven days incubation, weights ranged from 2.822 to 1.955 grams. The basic coloring of individual sets varied widely, often from a pale gray-white with a green tint to a pale rose-white. The spotting, concentrated on the large end, was distributed either as a fine splatter, as a broad mosaic of blotches, or as a combination of both. The red-brown coloration of the spotting also varied in intensity.

All but one egg observed through the hatching stage proved fertile, the exception being among a set of four.

Photograph by the auth[illegible]
February 7, 19[illegible]

BENT'S CROSSBILL

Three eggs is an average clutch. Note chicken feathers in the nest lining (Nest 2).

Incubation

From this study, it was concluded that incubation started primarily with the laying of the last egg. Though the females were seen to cover an incomplete clutch at night and during freezing temperatures, they often were absent during the warmer portions of the egg laying period. Also noted was the fact that, once the last egg arrived, the female rarely was found off the nest and then for the briefest of intervals, leaving only to defecate or at the insistence of the male bringing food. This was further strengthened by the fact that in all three nests observed from laying to hatching, the eggs of each set were pipped within a maximum period of ten hours, and all on the 14th day after the last egg had been laid.

The incubating female displays an amazing amount of bravery. Often she will not leave the nest until the observer approaches within arm's reach. Prior to hatching, she may be lifted off the nest and replaced at will. One female, having incubated only six days, let us band, weigh, and measure her, and accepted food readily from our fingers. In two instances, females on the nest were taken by predators, as was indicated by feathers found in deserted cavities.

On the other hand, the males were quite wary, rarely coming to the nest in the presence of an observer and then with much caution. At such times, the male would display considerable predator response, calling with great concern while skirting the nest area. If it were possible for him to call the female off, he would not approach the nest at all.

The male took no part in the incubation, but was responsible for the food supply. His deliveries often were irregular, ranging in frequency from once an hour to once every three hours. One extreme case, in which a male had not been seen to visit nest No. 6 all one afternoon, led to interesting speculation.

Despite the fact that the male had been observed aiding in the construction of nest No. 6, he had not been seen during early incubation. On March 6th, hoping to determine the male's function in this particular nesting, a five hour vigil was kept. Although a male was not seen at the nest, the female's behavior was noteworthy. Obviously inconvenienced by this lack of service, she was quite restless and called frequently to passing males occupied with their own domestic activities. Despite our offerings of food, which she accepted, this vocal activity continued until shortly after a male with two young birds in tow passed the area. She then left the nest in the direction the trio had taken, only to return in three minutes to resume incubation. On successive days, the male and youngsters were seen in the area

Photograph by the autho[r]
January 18, 195[illegible]

BENT'S CROSSBILL

The youngsters' craws are often literally stuffed with food (Nest 1).

and twice she left the nest in their company. Fortunately, due to banding, the trio was identified as being members of nest No. 7, from which the young had been fledged one day after construction on nest No. 6 had been started. This left us with two alternatives; first, that female No. 6 had lost her mate and had obtained the services of male No. 7, or second, that this was an example of double brooding, as mentioned by McCabe and McCabe (1933).

> A strong tradition of double-brooding, that is, two broods in rapid succession repeatedly crops up. It was categorically asserted for the genus by Bonaparte and Schlegel (1850), and has been vigorously advocated lately by Walpole-Bond and others, but without good evidence. Whitherby .... cites Yarrel, Jourdain, Dresser, and Seebohm as favoring such a belief, and to these should be added Kirkman ....

This alternative was strongly favored, not only because of the chronological sequence of the two nests, but also because we had never seen a female with the male and two juveniles, other than the bird from nest No. 6. Niedrach reported similar evidence in support of double-brooding from his 1947-1948 observations, when males with young were seen frequently, the females being unaccountably absent.

Nestlings

With the hatching of the eggs, the female brooded with strict regularity until the once naked youngsters sported a soft fluff of dark gray down. She then started to share with the male the responsibility of foraging for food, leaving the nest for longer and longer intervals. By the time the nestlings were well feathered, she covered them only at night and during severe weather. Just prior to fledging, covering of the young ceased entirely.

Both parents were quite busy during the nestling period, their rapidly growing young demanding more and more food. Following a night's fast, feeding visits to the nest were frequent. Later in the morning when the youngsters' craws were literally stuffed, the adults would come less often, taking time to forage for themselves. Each return to the nest was terminated by a careful house cleaning, the parents disposing of the fecal sacs either by swallowing them or by dropping them some distance away. As time passed, however, and the nestlings were able to discharge the sacs at the nest's edge, house cleaning often was neglected, the net result being a generous trim of white on the outer portions of the homestead.

As observed on Genesee Mountain, the diet proferred to the nestlings consisted mainly of sodden pine seed disgorged from the

BENT'S CROSSBILL

Wide open mouths welcome the food laden parents (Nest 1).

Photograph by the autl
January 18, 1

Photograph by the a
January 18,

BENT'S CROSSBILL

The male feeds the young while his mate begs for a share to pass on (Nest 1).

throats of the parents. With birds just hatched, however, the menu appeared dark and viscid, suggesting a composition of animal matter. Despite the lack of insect life in winter, the source of such a supply might be explained by the presence of caterpillar tents of the tiger moth, the inhabitants of which were active on warm days. Very likely, further search in the area would have revealed other sources in the form of dormant insect life.

The nestlings spend an average of 18 to 20 days in the nest from hatching to fledging. Following the development of down, the first pinfeathers appeared on the wings and tail on about the sixth day, the other feather tracts showing shortly thereafter. By the 14th day, they appeared well feathered, the head and back hoary with tufts of down adhering to the tips. By fledging time, the young had grown considerably, being six-sevenths of the average adult weight. Their wings and tails measured three-fifths as long as those of the adults, and their uncrossed beaks showed a little over one-half total development.

Fledging does not always terminate the active use of the nest. On February 23rd at site No. 3, the young popped while being observed, one youngster flying over a hundred feet and gaining altitude. The following day, all were back in the nest. This and similar observations at site No. 2 indicated that the young often return to the nest following their initial flight.

Once free of the nest, the young continued to be dependent upon the adults for food. While still mastering flight, the fledglings were observed to sit for hours chirping intermittently, thereby guiding the food-bearing parents to their hidden perches. Quick to spot approaching danger, they became silent and motionless, their streaked juvenile plumage serving well as camouflage among the pine needles and cones.

In a few days, however, the parent birds became continually harassed, the youngsters following their every move, getting in the way, demanding with open mouths and chirping excitedly when their efforts were rewarded with nourishment. After a week or so, the young became more adventuresome, going on ahead of the parents, or taking side excursions for short distances. At this time, they began to experiment with their slowly crossing beaks, picking tentatively at pieces of bark, cones or tufts of needles.

How long they are dependent on the adults is not known. A group of juvenile birds, many in postjuvenile molt, was observed in July, 1952 foraging quite well for themselves. However, measurements proved that their mandibles were still under adult proportions. On the other hand, at Genesee Mountain an identified pair was seen feeding the young 33 days after fledging. Somewhere between, emancipation occurs.

Nesting Mortality

The following table represents nests observed prior to hatching through to desertion either by fledging, predator action, or other natural causes.

TABLE No. 6

Nesting success of *Loxia c. benti.*

Location	Year	Nests	Eggs laid	Eggs per nest	Hatched no.	Hatched %	Fledged no.	Fledged %	Fledged per pair
Genesee Mt.	1952	11	33	3.0	25	75.7	11	33.3	1.0

Predators observed consisted of the Steller's jay (*Cyanocitta stelleri diademata*) and the tuft-eared squirrel (*Sciurus aberti ferreus*). The loss of eggs or small young was usually attributed to these, though not established. Crows were seen overhead, evoking predator response from adult crossbills. Several of the smaller hawks and owls were suspect due to the correctness of the ecology, but were not observed. In like manner the long-tailed weasel (*Mustela frenata nevadedensis*) and the ermine (*Mustela erminea muricus*) were believed to be present.

As observed by Bailey in 1948, heavy snows and high winds had little effect on nesting success. The maximum snowfall occurred on March 24, amounting to fourteen inches. Many nests were covered; the females, however, remained on the nests making access openings at the sides. Likewise sub-zero temperatures had little disrupting influence. The only instance of nestlings freezing was due to the failure of the female to brood during the night.

Three females were lost during the study. Two of these were killed at night on the nest presumably by owls or weasels. Two second nesting attempts were abandoned at approximately seven days incubation for reasons unknown.

Summary

Realizing that the Museum's observations might be peculiar to the Colorado race *benti*, a definite pattern of nesting has been observed during the past twenty years. They have been known to breed extensively on years coincident with plentiful coniferous seed supply, and it is possible that small nesting communities occur every year.

Photograph by A. Lang Baily
February, 1952

BENT'S CROSSBILL

Snowfall has little effect on the incubating females
(Nest 2).

There has been a decided shift of breeding through the ecological life zones on years the cycle has been initiated by any numbers. Usually commencing low in the foothills, the cycle has started in late December and slowly moved higher, reaching the Hudsonian Zone by midsummer. There the cycle continues as long as there is adequate food supply. The longest observed breeding residency of a particular group of birds in one life zone thus far has been from December through June in the Transition Zone.

Colorado *benti* plumage also has held to a general pattern. The majority of adult males have been observed as red, and the majority of postjuveniles as multicolored, with a scattering of exceptions. Yellow and particolored males have been noted as rare. The principal annual molt occurs, irrespective of breeding activity, for both juveniles and adults between June and October, the height being July through August.

The use of the crossed mandibles was carefully observed and notes taken on feeding habits. Noteworthy was a breeding male taken in July feeding entirely on aphis, and the behavior of securing scrapings from bleached deer bones.

The character of the transitory flock was seen to change as the breeding cycle started, with the birds forming a loose knit community in favorable areas. The density of population within communities varied with the local supply of food. Foraging flocks of breeding birds were often formed and were frequently found in conjunction with winter association groups of other species.

Pre-nuptial activity was characterized by the male's nuptial flight and song, as well as his nuptial feeding of the female, the latter behavior continuing throughout the pre-egg period. Territorialism was observed only at the time of nest site selection, being apparently disregarded thereafter.

Nest activity lasted from 43 to 48 days from start of construction to fledging; nest construction, 5 days; completion of nest to first egg, 4-5 days; egg laying at one per day, 2-4 days; incubation, 14 days; and nestling period, 18-20 days. The study was unable to determine if there was variation in this pattern at higher altitudes during summer months.

The female was seen to assume the leading role in nest site selection and construction. Choice of nest sites varied, though all were located in yellow pine, averaging 20 to 40 feet above the ground. The majority were found on the southern sides of the trees. Those nests located terminally were far more successful than those on lateral limbs. Material used in nest construction was consistently similar, the variations reflecting the immediate nest area.

Nest No.	Tree girth chest high (inches)	Tree height (feet)	Nest height (feet)	Slope exposure	Nest limb exposure	Nest exposure	Nest to tip of limb (feet)	Date found	Condition when found	Nest finished	First egg	Last egg	Hatched	Fledged	No. fledged
1	42″	50′	26′	S	S	S	3′	Jan. 16	3N	....	....	*female failed to brood on night of Jan. 18, young frozen*			
2	31″	26′	24′	S	T	E&W	2′	Jan. 26	SS	Jan. 31	Feb. 4	Feb. 6	Feb. 19	Mar. 9	3
3	49″	43′	40′	N	T	SE	3′	Feb. 2	3E	....	....	....	Feb. 3	Feb. 23	3
4	58″	55′	18′	NE	S	S	4′	Feb. 11	3E	....	....	....	Feb. 18	*nest empty on Mar. 1*	
5	......	50′	40′	SW	SE	SE	3′	Feb. 11	C	*nest abandoned ½ completed*					
6	49″	38′	18′	NE	NW	SW	2′	Feb. 15	SC	Feb. 20	Feb. 25	Feb. 28	Mar. 13	*nest empty on Mar. 25*	
7	45″	44′	15′	NE	SE	S	3′	Feb. 16	2N	....	....	....	....	Feb. 16	2
8A	63″	60′	46′	SE	SE	S	3′	Feb. 20	C	....	Feb. 23	Feb. 25	*nest empty on Feb. 27*		
8B	70″	73′	30′	SE	SE	S	5′	Feb. 28	SS	Mar. 6	Mar. 10	Mar. 12	*nest empty on Mar. 19*		
9	49″	48′	44′	E	T	E	4′	Feb. 23	4E	....	....	....	Mar. 1	Mar. 20	3
10	47″	40′	20′	NE	N	N	5′	Feb. 23	C	Feb. 28	Mar. 3	Mar. 5	Mar. 19	*nest empty on Mar. 28*	
11	48″	50′	35′	S	S	SW	2′	Mar. 5	C	Mar. 9	*female on 2 eggs Apr. 7, abandoned Apr. 9*				
12	61″	60′	48′	N	S	S	3′	Mar. 19	....	*nest inaccessible*					
13	50″	47′	20′	SW	E	NE	2′	Mar. 16	SS	Mar. 25	Mar. 30	Mar. 31	*nest abandoned Apr. 8*		
14	49″	45′	32′	SW	S	SE	3′	June 3	1N	....	....	....	....	June 6	1
15	58″	55′	40′	NE	S	S	4′	June 3	4E	*nest and birds collected June 4, incubation fresh*					

Key to symbols.. C—under construction; SC—start of construction; SS—site selection; T—terminal growth; 3E—three eggs etc.; 3N—three nestlings, etc.

Photograph by the authors
June 4, 1952

BENT'S CROSSBILL

Set of four eggs found in early June (Nest 15)

Though located in yellow pines, no material was used in construction from that species.

Egg sets averaged three in number, varying to two and four. Only one infertile egg was observed. Incubation was established as starting with deposition of the last egg, since the females did not maintain a rigid incubating schedule until that time and all sets observed hatched within 10 hr. periods. At no time were males observed to incubate.

Of significance was an instance that strongly indicated double brooding, and was felt to be the best evidence of such for the species as yet recorded. Also of interest was a case where a nest served for a second set of eggs following the destruction of the first.

The young hatched almost naked, but quickly developed down, following which the female began to aid the male in gathering food. After the fourteenth day all feather tracts were well started, and later during fledging the young proved to be approximately 1/3 under total adult development. The nest may be used by the young several days after their initial flight.

Nesting success during the cycle averaged one bird fledged for every three eggs laid. Three nesting females were recorded as lost to predators.

Bibliography

AIKEN, CHARLES E. H and EDWARD R. WARREN

1914 The Birds of El Paso County, Colorado: Colo. College Science Series, 12:455-603.

ALDRICH, ELMER C.

1939 Notes on the Salt Feeding Habits of the Red Crossbill: Condor, 41:172-173.

1940 Notes on the Birds of Crater Lake National Park, Oregon: Condor, 42:89-90.

ALEXANDER, GORDON

1937 The Birds of Boulder County, Colorado: Univ. of Colorado Studies, 24:79-105.

ALLEN, AMELIA S.

1920 Red Crossbills at Berkeley, California: Condor, 22:73.

ALLEN, J A. and WILLIAM BREWSTER

1883 A List of Birds Observed in the Vicinity of Colorado Springs, Colorado: During March, April, and May, 1882: Bull. Nutt. Orn. Club, 8:151-61, 189-198.

BERGTOLD, WILLIAM H.

1917 The Birds of Denver: Wilson Bull., 29:113-129.

1919 The Season: Bird-Lore, 21:370.

1928 A Guide to the Birds of Colorado: Smith-Brooks Printing Co., Denver, Colorado, 270 pp.

BETTS, NORMAN D.

1913 Birds of Boulder County, Colorado: Univ. of Colorado Studies, 10:177-233.

BLACK, J. D.

1932 The Red Crossbill at Lawrence, Kansas: Auk 49:489-490.

BRENINGER, GEORGE E.

1894 American and Mexican Crossbills: Nidologist, 1:99-101.

BREWSTER, WILLIAM

1886 The Red Crossbill (*Loxia curvirostra stricklandi*) in Kansas: Auk, 3:260.

1918 Nesting of the Red Crossbill (*Loxia curvirostra minor*) in Essex Co., Mass.: Auk 35:225.

BROOKS, MAURICE

1942 White-winged Crossbills and Sitka Crossbills summering in West Virginia: Auk 59:118-119.

COOKE, W. W.

1897 The Birds of Colorado: Bull. No. 37, Tech. Series No. 2, State Agric. College of Colo., 143 pp.

1900 The Birds of Colorado, Second Append. to Bull. No. 37: Bull. No. 56, Tech. Series No. 5, State Agric. College of Colorado, pp 179-239.

COUES, ELLIOT

1874 Birds of the Northwest: Misc. Pub. No. 3, U.S. Geol. Survey of the Territories: F. V. Hayden, U.S. Geologist in Charge, Washington, D. C., 791 pp.

DREW, F. M.
1881 Field Notes on the Birds of San Juan County, Colorado: Bull. Nutt. Orn. Club, 6:138-143.

DYCHE, L. L.
1886 The Red Crossbill (*Loxia c. stricklandi*) in Kansas: Auk, 3:258-261.

EMERSON, W. OTTO
1899 American Crossbills in Alameda County, California: Condor, 1:30.

FISHER, WILLIAM HUBBELL
1888 The American Crossbill, as to some of its Habits and its Fondness for Salt: Jour. Cincinnati Soc. Nat. Hist., 10:203-204.

GILMAN, M. F.
1907 Some Birds of Southwest Colorado: Condor, 9:152-158.

GRISCOM, LUDLOW
1937 A Monographic Study of the Red Crossbill: Proc. Boston Soc. Nat. Hist., 41:77-210.

HAYWARD, C. LYNN
1943 Notes on the Status of the Red Crossbill in Utah: Auk, 60:276-278.

HENDERSON, JUNIUS
1903 Preliminary List of Birds of Boulder County, Colorado: Univ. of Colorado Studies, 1:233-237.
1907 Nesting of Crossbills in Colorado: Auk, 24:440-442.
1909 An Annotated List of the Birds of Boulder County, Colorado: Univ. of Colorado Studies, 6:219-242.

HENSHAW, HENRY W.
1875 Report upon Ornithological Specimens Collected in the Years 1871, 1872, 1873; Engineering Dept. U. S. Army, Geographical and Geological Explorations and Surveys West of the One Hundredth Meridian: Govt. Printing Off., Washington, D. C., 5:133-507.

HERSEY, LUMAN J. and ROBERT B. ROCKWELL
1909 An Annotated List of the Birds of Barr Lake District, Adams County, Colorado: Condor, 11:109-122.

HOFFMAN, RALPH
1926 Male American Crossbill feeds Female: Condor, 28:48.

JOLLIE, M.
1953 Plumages, Molt, and Racial Status of Red Crossbills in Northern Idaho: Condor, 55:193-197.

KELLOGG, VERNON L.
1890 Summer Birds of Estes Park, Colorado: Trans. Kan. Acad. Science, 12:80-90.

KLEINSCHNITZ, FERD C.
1937 Field Manual of Birds of the Rocky Mountain National Park: Private Printing, Denver, Colo., 60 pp.

LINCOLN, FREDERICK C.
1915 Birds of Yuma County, Colorado: Proc. Colo. Museum of Nat. Hist., December, 10 pp.

MACPHERSON, H. A.
1889 Crossbill breeding in Immature Plumage: Zoologist, 13:229.
MARSHALL, WILLIAM H.
1940 More Notes on Salt Feeding of Red Crossbills: Condor. 42:218-219.
McCABE, THOMAS T. and ELINOR B. McCABE
1933 Notes on the Anatomy and Breeding Habits of Red Crossbills: Condor, 35:136-144.
MEADE, GORDON M.
1942 $CaCl_2$, A Death Lure for Crossbills· Auk, 59:439.
MERRILL, J. C.
1898 Notes on the Birds of Fort Sherman, Idaho: Auk, 15:14-22.
MILLER, CHRISTY
1914 On the Crossing of the Bill of the Crossbill: British Birds, 7: No. 11.
MILLER, LOYE
1920 Notes from the Region of Lake Tahoe: Condor, 22:78-79.
MINOT, HENRY DAVIS
1880 Notes on Colorado Birds: Bull. Nutt. Orn. Club, 5:223-232.
MORRISON, CHARLES F.
1887 Field Notes on some Birds of Colorado: Orn. and Ool., 12:35.
1888 A List of Some Birds of La Plata County, Col., with Annotations: Orn. and Ool., 13:70-75.
1889 A List of the Birds of Colorado: Orn. and Ool., 14:148-149.
MUNRO, J. A.
1919 Notes on the Breeding Habits of the Red Crossbill in the Okanagan Valley, British Columbia: Condor, 21:57-60.
NIEDRACH, ROBERT J. and ROBERT B. ROCKWELL
1939 The Birds of Denver and Mountain Parks: Colo. Museum of Nat. Hist., Popular Series No. 5, 196 pp.
PACKARD, FRED MALLERY
1945 The Birds of Rocky Mountain National Park: Auk, 62:371-394.
PEABODY, REV. P. B.
1907 The Crossbills of Northeastern Wyoming: Auk, 24:271-278.
PHILLIPS, JOHN C.
1912 The Hawaiian Linnet, *Carpodacus mutans* Grinnell: Auk, 29:336-338.
POLING, OTHO C.
1890 Notes on the Fringillidae of Western Illinois: Auk, 7:238-243.
RACEY, KENNETH
1948 Birds of the Alta Lake Region, British Columbia: Auk, 65:283-401.
RICHARDSON, FRANK
1938 Red Crossbills feeding on Juniper Galls: Condor, 40:257.
RIDGWAY, ROBERT
1883 A Review of the American Crossbills (*Loxia*) of the *L. curvirostra* Type: Proc Biol. Soc. of Wash., 2:101-107.
ROBBINS, C. A.
1932 The Advantages of Crossed Mandibles: A Note on the American Crossbill: Auk, 49:159-165.

ROCKWELL, ROBERT B.

1908 An Annotated List of the Birds of Mesa County, Colorado: Condor, 10:152-180.

ROCKWELL, ROBERT B. and ALEXANDER WETMORE

1914 A List of Birds from the Vicinity of Golden, Colorado: Auk, 31:309-333.

ROSS, EDNA G., and VERNA M. ROSS

1950 Nesting of the Red Crossbill in Pakenham Township, Lanark County, Ontario: Canad. Field-Nat., 64:32-34.

SCLATER, WILLIAM LUTLEY

1912 A History of the Birds of Colorado: Witherby and Co., London, xxiv, 576 pp.

SCOTT, W. E. D.

1879 Notes on Birds observed at Twin Lakes, Lake County, Colorado: Bull. Nutt. Orn. Club, 4:90-95.

SELANDER, ROBERT K.

1953 Notes on the Red Crossbills of the Uinta and Wasatch Mountains, Utah: Condor, 55:158-159.

SILLOWAY, P. M.

1905 The American Crossbill in Montana: Condor, 7:174-176.

SNYDER, DANA PAUL

1950 Bird Communities in the Coniferous Forest Biome: Condor, 52:17-27.

SNYDER, DANA PAUL and J. FRANK CASSEL

1951 A Late Summer Nest of the Red Crossbill in Colorado: Wilson Bull., 63:177-180.

SNYDER, DOROTHY E.

1952 Red Crossbills: Bull. Mass. Aud. Soc., 36:383-386.

STORER, TRACY I.

1921 American Crossbill eating Elm Aphis: Condor, 23:98.

SUTTON, GEORGE M.

1943 Records from the Tucson Region of Arizona: Auk, 60:345-350.

SWENK, MYRON H.

1921 Nesting of the Red Crossbill in Nebraska: Wilson Bull., 33:38-39.

TANDY, E.

1889 Assumption of the Male Plumage by the Female Crossbill: Zoologist, 13:182.

TICEHURST, C. B.

1915 Plumages of the Male Crossbill, *Loxia curvirostra*: Ibis 3:662-669.

TORDOFF, HARRISON B.

1952 Notes on Plumages, Molts, and Age Variation of the Red Crossbill: Condor, 54:200-203.

TRAVERNER, P. A.

1922 Crossbills eating Aphis: Condor, 24:96.

WARREN, EDWARD E.

1910 Bird Notes from Salida, Chaffee County, Colorado: Auk, 27: 142-151.

1916 Notes on the Birds of the Elk Mountain Region, Gunnison County, Colorado: Auk, 28:292-317.

WIDMAN, OTTO

1911 A List of Birds observed in Estes Park, from June 10, to July 18, 1910: Auk, 28:304-319.

To illustrate, in the Atchison, Topeka & Santa Fe reorganization of 1895, certain bondholders were given $75 in new mortgage bonds and $40 in income bonds—a total of $115 in securities—for each $100 of their old mortgage bonds, the interest payment on the income bond being conditioned "if earned." As these bondholders had received no interest for three years prior to the reorganization, they were (in effect) to be paid defaulted interest later **if future operations proved profitable.**

ADD PAGE HERE

zation, has been more widely criticised than any other one feature of railroad financing. A fair statement of the conditions obtaining at the time of reorganization would be, that the adjustment, as between the various security holders, has usually been fair, and that the increased capitalization represents previously defaulted interest on the bonds and new cash put into the property for the purpose of rehabilitating it, paying the floating debt and providing a working capital.

Printed in Dunstable, United Kingdom

91220745R00038